FROM **OIL** TO **GAS**

by Shannon Zemlicka

Lerner Publications Company / Minneapolis

Lerner Publications Company
A division of Lerner Publishing Group
241 First Avenue North
Minneapolis, MN 55401 USA

Website address: www.lernerbooks.com

Library of Congress Cataloging-in-Publication Data

Zemlicka, Shannon.
    From oil to gas / by Shannon Zemlicka.
        p.   cm. — (Start to finish)
    Includes index.
    ISBN: 0–8225–0718–8
        1. Petroleum—Juvenile literature. [1. Petroleum industry and trade.] I. Title. II. Start to finish (Minneapolis, Minn.)
    TN870.3 .Z45 2003
    338.7'6223382—dc21   2001007215

Manufactured in the United States of America
    2 3 4 5 6 – JR – 08 07 06 05 04 03

The photographs in this book appear courtesy of:
© Gregg Otto/Visuals Unlimited, cover, pp. 1 (bottom), 23; © Howard Ande, pp. 1 (top), 3, 11, 21; © Inga Spence/Visuals Unlimited, p. 5; © Ron Sherman, pp. 7, 13; © Lowell Georgia/CORBIS, p. 9; © TRIP/H. Rogers, p. 15; © TRIP/B. Turner, p. 17; © TRIP/TRIP, p. 19.

Table of Contents

Gas makes cars go.

How is it made?

Workers look at the land.

Gas comes from a thick, black liquid called **oil.** Oil is found deep underground or under the ocean. It must be dug up. On land, workers choose where to dig by looking for clues in the land.

Workers clear the land.

Workers bring bulldozers to land that may have oil. The bulldozers clear away trees and bushes.

Trucks bring tools for digging up oil.

Trucks bring tools to the cleared land. One tool is a machine that makes electricity. It powers a huge drill for digging. Another machine can pump up mud and rock from underground.

Workers put the tools together.

Workers put up a large frame. Then they add the machines to the frame. The frame holds the machines together to make a **rig.**

The drill digs for oil.

A worker runs the rig's drill. The drill digs a deep hole called a **well.** Some wells have no oil. The workers move the rig and try again until they find oil.

Pipes carry the oil.

The oil rushes up into the well.
Machines pump the oil into pipes.
The pipes carry the oil to huge
tanks.

The oil goes to a factory.

Trucks, trains, ships, and long pipes carry the oil to a factory. The factory is called a **refinery.** A refinery looks like a maze of tanks, towers, and pipes.

Pipes heat the oil.

Hot pipes heat the oil. Then it enters a tower. The heat separates the oil into many parts. Gas is one of these parts.

The gas goes to gas stations.

The gas travels to gas stations in trains, trucks, or pipes. Workers store the gas in large tanks underground.

Fill it up!

At gas stations, drivers use pumps to fill their cars and trucks with gas. Then it's time to get back on the road!

Glossary

oil (OYL): a thick, black liquid found underground

refinery (ree-FY-nuh-ree): a factory where gas is taken out of oil

rig (RIHG): a frame and machines that are used to dig up oil

well (WEHL): a hole that is dug to find oil

Index